Why Farm

Selected Essays & Editorials

L. R. Miller

Why Farm, Copyright 1997 Lynn R. Miller

Publisher
Small Farmer's Journal Inc.
PO Box 1627
325 Barclay Drive
Sisters, Oregon 97759
541-549-2064

Printed in the United States of America
Printer: Parton Press, Redmond, Oregon

Authored by Lynn R. Miller

First Edition,
First printing 1997

ISBN 1-885210-05-1

Also by L. R. Miller

The Work Horse Handbook
Training Workhorses / Training Teamsters
Buying and Setting Up Your Small Farm or Ranch
Ten Acres Enough: The Small Farm Dream Is Possible
Thought Small: Poems, Prayers, Drawings & Postings

*This book is dedicated to the
memory of my son
Ian Lewis Miller
who, in his short lifetime,
always knew why*

preface

Since 1976 I have served as the editor of **Small Farmer's Journal**, *an international underground agrarian quarterly I established in that same year. I was not given the job of editor but rather took it because no one else wanted it. I was not qualified for the work but the cause was too great to allow such an obvious condition to slow my naive rush to battle. I was blinded to my inadequacies by my passion. And any skill with editing and writing, which I might have developed over time, has been diluted by the fact that my best moments are saved for the work I do on my small ranch and the time I spend with my family.*

For twenty two years, in my questionable capacity as editor, I have written dozens of editorials, essays, and articles which have, over time, given shape to a individualistic philosophy centered around a modern agrarian dream. This book is a compilation of some of my earliest writings revised only slightly for presentation today.

These writings are presented in book form not because they are good but rather because the cause they serve is important and the ideas, however crudely presented, may have some constructive value.

I believe in the repopulation of the farmscape. I believe in the intelligent, romantic, independent diversified small family farm.

Lynn R. Miller 1997

9

Contents

Chapter One

Thoughts on Farming

Wе all live by farming, regardless of whether we live in cities or in the country. Our bodies spring from the earth. We are all sustained by a fragile layer of decaying organic matter and mineral substances we refer to as top soil. Top soil, water, and sunlight give plants life and plants, in turn, give life to us. We live in a fantastic complex unified system. If you trace down all the leads other life becomes our own. And our own life becomes other life. In death we return to the earth. Beyond the chemistry there is continual mystery and all the other ingredients of religion.

We have come to ignore, disregard and violate what for centuries had been a sacred trust - the stewardship of the land. The last seventy years have seen an incredible industrialization of agriculture in this country. Machinery and chemicals have combined to give us an agriculture which can and does boast of the highest productivity per farmer of any country in the world. This is misleading. We do not enjoy the highest productivity per

This material, in slightly altered form, originally appeared in Vol. 2, No. 3 (1977) of Small Farmer's Journal.

acre or the greatest efficiency of production. The technology that belongs to this new orthodoxy of farming is designed for extensive monocultural practices, thousands of acres of grain or many hundreds of acres of a given row crop.

Before the advent of this industrialization of farming, the nature of food supply systems required diversity in order to provide for regional needs. A diversity of all aspects of farming coupled with the large number of farmers and farm workers accounted for a rich social fabric. There are a great many people today who, with or without first-hand experience, sincerely believe that the quality of life then was superior to what we have now. Proponents of our modern agriculture argue that we cannot turn the clock back and that these people are forgetting or never knew the inherent hardships of our old order of farming. They argue that to return to such a system of human scale farms practicing diversified agriculture and relying primarily on regional markets would cause wholesale famine and chaos. They argue that the realities of our present day economy would force on all of us a much lower standard of living (measured by dollars and convenience) if such a system came to be. They argue that this change would disrupt the existing agricultural support industries and trade relationships resulting in a wholesale economic depression.

Public understanding of the whole question is confused because so much of what is read and heard is born of incomplete or incorrect logic and data.

Before addressing myself to the specific arguments and

concerns involved, I would like to go into the structure of our new orthodox agriculture as I understand it.

First of all, I do believe that farming in this country is in sad shape and there is a critical vulnerability within this structure.

It was perhaps inevitable that our agriculture should evolve to its present state considering the atmosphere of big government and big business that has existed since early in the twentieth century. Many drastic economic and technological changes in farming were directly related to official responses to the economic depressions and wars in which this country has been involved. Government and big business including banking sought to insulate themselves from the effects of depressions in the economy by controlling production, distribution and wages thereby affecting supply and demand in the market place. Whether or not this was a proper approach theoretically is not as important as the fact that these sorts of controls continue to be implemented without cooperation or organization between the powers involved. So we have a messy disorganized government and a crazy, confused industry-labor community. The only common goal seems to be the continual desire for economic growth. As obvious as it seems, the correlation between this growth and inflation is ignored. Because of the unchecked productivity that comes of supplying a war effort, we experience tremendous growth in terms of new manufacturing facilities, commodities and jobs. After the war this growth must be redirected or a depression will follow. Agriculture has absorbed a great amount of this growth after each war. We did not see wholesale application of herbicides and pesticides in farming until after WWII when the companies which manufactured poisonous chemicals to defoliate the islands of the South Pacific and control disease carrying insects redirected to supply farmers with not only these substances but also incredible propaganda selling chemical controls as the saving grace of agriculture. Many other examples could be cited illustrating the effect of wars and economic controls on the shape of agriculture as we know it today.

So agriculture has evolved painfully to its present disorganized structure due to the outside pressures of government and big business whose interest is not in the internal workings of the farm but rather in the control of products, productivity and purchased inputs such as machinery and chemicals. It is interesting to imagine how agriculture may have evolved had its evolution been controlled by farmers, a free market and the weather. There are some good examples of this in other countries.

Orthodox agriculture today is a world of monocultural cropping, expensive specialized machinery, enormous surpluses in certain commodities and extensive land holdings.

The economics of our modern agriculture are fragile indeed. We've heard a great deal about the plight of the farm with this farmer's strike business. Most all of the active members of this movement are farmers with large holdings who grow grains, soybeans or other crops and must rely on the bulk sale of a single commodity at the lowest wholesale price during a period of the year when most other farmers have their like crop for sale.

These farmers have invested heavily in a system and structure
of agriculture that was sold or force fed on them by the government, USDA, Extension Agents, farm banks, manufacturers and
farm magazines. These farmers have borrowed enormous sums
of money on short term for high interest to purchase ever newer
equipment, ever more expensive chemicals and for general
operation. They are convinced by the insanity of liberal economics which suggests that the only way to get out of the hole is to
borrow more money and expand operations.

It is understandable that these farmers who are suffer-
ing from not only substantial financial losses but also uncertainty
about future markets should turn to the government for guaran-
teed prices   for their products. These farms are fighting for
their survival. It is unfortunate and paradoxical that they are
convinced that the system of agriculture which has made their
existence precarious is a correct system. They will not know
any measure of continued security until they are able to break
away from the economic constraints of this system.

While there is much talk about the low prices that
farmers get for their goods we also hear complaints about the
high cost of food to the consumer. Within our current structure
of production and distribution there are some gross inequities.

In 1975 government figures indicated that 42 cents of
each consumer dollar spent for food went to the farmer and 58
cents went to so-called middlemen. We as consumers are a
convenience oriented group. Many of us would prefer not to

peel potatoes. At the same time a lot of us are now enjoying the luxury of a concern for health and the quality of our food. There were those 15 years ago who excitedly predicted that the health food craze would have many households enjoying whole natural foods such as the potato. They talked of imminent doom for the processed food industry. Yet you can today go to most health food stores and buy natural potato chips and a wide variety of other organic prepackaged processed foods.

In a recent newspaper article Rep. Eckhardt of Texas was quoted defending food retailers as only No. 27 among the top 30 industries in America in terms of profitability and therefore not responsible for high food prices. Rep. Eckhardt accused food processors and distributors of adding cost without adding value. Excessive advertising, packaging, unnecessary additives and aggressive marketing of so-called convenience, combined with a production and marketing system dependent on high-cost energy and transportation and inadequate competition were cited as the areas responsible for food prices being where they are.

An advertising specialist R.A. Baskin was quoted in the same article as saying that the nation's food advertising bill is now about $5 billion a year and that that is a cost that comes off the top of everybody's food budget. Jacquelyn Helm, director of United We Resist Additional Packaging (UNWRAP) claims that food packaging is now a $23 billion a year industry and that does not include the cost of disposing of the 55 million tons of solid waste created by used packages. Eight to 12 percent of the family's food budget is now spent on packaging.

Rep. Eckhardt noted that food wholesaling and food processing are respectively the fifth and ninth most profitable industries in the US.

Meanwhile, back at the farm, the wholesale industrialization of agriculture has made food production critically dependent on petroleum products. In 1974 farming consumed 16.5% of the total energy used in this country. That was up from 10 percent in 1970. USDA figures gleaned from the federal census in 1974 show that the total farm-related use of gasoline accounted for 3,683,822,000 gallons, diesel 2,618,548,000, fuel oil 304,342,000, natural gas 173,221,376,000 cu. ft. and electricity 32,344,608,000 kilowatt hours.

Some of our more intelligent ag-economists believe that we can no longer afford to measure the efficiency of our agricultural production in terms of the dollar cost as versus the wholesale value. Considering that fossil fuels are nonrenewable resources and they figure prominently into the total cost of American agriculture, it would be appropriate to figure energy into the balance sheet.

Farming produces calories. By caloric measurement of energy it takes us now 13 to 14 petroleum calories to produce one food calorie. In this context we have the world's worst system of agriculture. If the farmer had to pay out 13 to 14 dollars to produce a dollar's worth of food he would go out of business. At the rate we are depleting our petroleum resources our system of agriculture is short-lived. I maintain that as it is bankrupt not only by caloric efficiency measure but also morally. But I'll save that for a little later.

This agriculture we have now has progressed from a growing vulnerability to a dangerously precarious situation. If the relatively few big farmers (4% of the total) who supply through interstate trade 50% of our food continue to go out of business or are wiped out by drought or foreign food stuffs sold in this country who could measure the net effect to all of us? If government in its own meddling and incompetent way chooses to get further involved by price supports, loans and production cut-backs it is terrifying to imagine the garbled mess that would result. If the supply of petroleum products were to be cut back substantially how would the farmer operate? We've got a lousy system now but what could come of it in the near future far outshadows this mess. We've got very little to fall back on.

Now for those of you who think I'm about to fall back into nostalgic reverie and talk about going back, you're wrong.

There are options and certainly plenty of good present day models. Perhaps some far-sighted sensitive communities will implement constructive change. Perhaps some destructive patterns can be halted. Perhaps not.

But as for the original rebuttal I promised to stock-responses of the proponents of orthodox agriculture: First off I know of no intelligent person, concerned about our agriculture, who advocates a return to the farm of 1900 as a model for social salvation. I believe that what I know from my father and grand-father and from extensive studies of farm history indicates that the quality of life in many instances was superior to what we have now. There were exceptions but I'm interested in the positive examples and what they can suggest to us now for constructive change. Of course we can't turn the clock back. Yet it is ridiculous and short-sighted for us not to constantly re-examine our environment and use the important lessons of the near and distant past towards a better future. I for one am arguing for a more constructive evolution to our farming for a better world tomorrow.

We can change to a system which identifies and respects the viability of small farms as important ingredients in the broad agricultural community. Efficient smaller farms do produce more per acre. A system of human scale (instead of industrial scale) agriculture practicing diversified mixed crop and livestock farming and relying to a large measure upon regional, direct and semi-direct, marketing would not cause famine or chaos. To imagine that such a change were to happen overnight is a ridiculous assumption made by the proponents of orthodox agriculture to bolster the scare tactics of their arguments. More than enough dedicated people with a sincere devotion to farming, and a desire to be farmers, are available today. Many people who have been forced out of agriculture by the current structure (not by their inefficiency or incompetence) are imminently qualified to return to a receptive system. Many more people with the dedica-tion but without working experience would leap at the opportunity to invest in their own education in farming if institutions existed with constructive and appropriate curriculum to suit their needs and wishes.

As to the question of a reduced standard of living:
Wendell Berry asks

*"are we failing to consider that a family might farm a
small acreage, take excellent care of it, make a decent,
honorable and independent living from it and yet fail to
make what the rest of us would consider a profit?"*

Our standard of living is a sore point because it is
misleading. We spend so much time and money trying to enjoy
ourselves, keep ourselves physically fit, and fill in spare time. I
maintain that our standard of living and the quality of our life are
not parallel at all. There are too many of us looking for something
better.

A change in our system of agriculture, to what I've
alluded, would not result in a depression because it would replace
jobs with jobs and because it could strengthen regional econo-
mies.

In our society today we have a confused set of values.
Far too many people have nothing to believe in. So it is easy to
see the correlation between our attitudes towards ourselves and
our attitude towards our environment. Very few people work
where they live. Without this it is difficult to develop a true
respect and appreciation for either environment.

There is recent official concern for what is referred to as
the small farm condition in this country. This is good and long
overdue even if it is misguided. Public and private institutions
have become convinced that the preservation of the small farm is

important to the total rural community. Programs are on the drawing board to provide for a whole new welfare system directed to the small farmer. This, coupled with the refusal on the part of these institutions to accept the viability of small farms in the economics of orthodox agriculture, is the reason why I say the concern is misguided. We do not need an elaborate welfare system for "marginal" farmers. We need for government to back out of agriculture altogether. But the concern is good because of the excellent dialogue it is generating.

In discussions of this sort polarities seem to be inevitable. Some people are on one side of the line and the rest on the other. But people are the issue not the problem. There are good people in government, in the extension service, owning and operating large farms, and making big business decisions. I must presume that they believe they are not only right but know where they're going. For the sake of a presentation such as this, with space a consideration, it seems impossible to make a point without making some arbitrary absolutes. I believe there is no only way to do anything. What's right for me may not be right for you. And I don't enjoy the either/or approach. Perhaps the best solutions come from constructive compromises.

My over-riding concern is that we consider farming from a moral standpoint. Our best values must be taken to the task of re-examining farming. The state of our farming has much to do with the nature of our values.

Chapter Two

Why Farm

G enerations ago what you did as your life's work was often dictated by family circumstance, cultural inheritance and education. In some special situations those are still determining factors. But for most of us cultural inheritance is too far removed, education has become a fairly common demoninator with less effect, and family circumstance has less bearing. Peer pressure and the unpredictable reactions to incredibly confusing media blitz's appear to have more influence over personal choices about vocation than traditional criteria. But personal preferences do not the end make. Wanting to be a movie star has more often created a restaurant worker than an actor. Glamour and money seem the most prevalent goals for young adults today. And as such, it is indicative of the cancer eating at our contemporary culture. If a life's work returns comfort, an abiding sense of purpose above and beyond personal gain, the satisfaction of small accomplishments made in close partnership with natural processes and a gentle maintenance of balanced humility - then that person has experienced a lifetime more complete than most. To achieve satisfaction, a sense of self-worth and completeness in a lifetime; these are a set of goals that few people can conceive of yet little else measures up to the full potential for humankind. These are honorable goals which, on the surface, may appear selfish but which in fact require service to others and our environs. Deliberately choosing a vocation which might provide the opportunity to reach these goals may be difficult. Some have been so fortunate as to "know", intuitively, from day-one what their life's work was to be. And the measure to which that "knowing" consumed them with a single-mindedness of purpose often contributed to the success of their great adventure. If you want to do something and work hard at it and enjoy the process it is difficult not to measure the time spent as a joyous success (no matter the gain). But if you don't have a real clue as to your life's path or have inklings of interests but no form to contain them, selecting a vocation will be difficult. Perhaps this discussion may help. As you may have guessed I'm about to suggest you take up farming as your life's work. I'm also going to remind some of you why farming is a path to re-examine, to stick with and perhaps to realign.

Farming for Art's Sake

Among those who might read this there are two groups

to which it is specifically addressed: one contains those who are yet to choose their work and might have an inkling that farming should figure in, the other contains those who have had very tough times in recent years and have either decided to quit farming or might be wondering why they ever got into it.

In a discussion entitled 'Why Farm?' it would be far too easy to list the bleak vocational alternatives to farming and talk about the difficulties in finding satisfying work that pays well and offers some semblance of job security. It would also be easy to point out the unpleasantness that often comes from the pressures of working for others, especially large impersonal corporations. But talking about those things takes us in the wrong direction. Instead of looking over our shoulders at what we might not want in our lives, it makes more sense to look ahead at what we would like to have in our lives and what we need our lives to mean.

Farming as a vocation is more of a way of living than of making a living. Farming at its best is an Art, at its worst it is an industry. Farming can be an Art because it allows at every juncture for the farmer to create form from his or her vision. A farmer sees, in the mind's eye, that the fields might look, feel and produce a certain way. Then all manner of tangible and intangible ingredients are brought into the creative process to bring those fields into that form. It is a complex molding and shaping process that involves four dimensions. It is a form of living sculpture with the fourth dimension being a knowledge that today's form has immeasurable effect on tomorrow's. As pure industry there is no value given to perceived (felt) satisfactions of progress and result. As Art those satisfactions are a major reward and motivation to continue. As industry the fourth dimension of anticipated long term effect has no value. As Art the future shape of the farming adventure keeps the spirit and hope alive and vibrant. Farming as a way of living contributes to the strength of the family structure and to the culture of local communities. As industry, farming cannot be concerned for these things because there are no immediate financial rewards to be had. The attraction to farming has, more often than not, been that it promises a humane way of living.

The Promise of Farming

It has been said often that Farming requires skills and aptitude in many areas, be they plant-science, mechanics, animal husbandry, marketing, engineering, etc. That means that Farming is a challenging occupation but certainly in more ways than applied science suggests. There are challenges of the spirit which come with the uncontrollable weather related failures of crops. There are challenges to mental well being and motivation when, as has happened recently, banks and government programs deem you unworthy. And both of the above puts pressure to bear on family relations. There are many forks in the road where a specific occurrence casts a shadow over the farming experience.

Then Why Farm?

Humans thrive on challenges both physical and spiritual and especially if there is the slightest promise of success. And in farming the PROMISE is for a noble comfort, a growing sense of purpose, a tangible sense of self-worth and the possibility of a continuity. No where in there will you find any mention of a promise of monetary wealth, fame or political power. Farming as a vocation has seldom succumbed to the manipulations, maneuvers and deceptions often called upon in the pursuit of fame and fortune. That is because it is impossible to separate out the less tangible spiritual elements of the creative process which have such an important part in the complex growing organism that is a farm. Yet, with the full knowledge that farming is a difficult business with little real opportunity for substantial cash income, thousands of people each year make the choice to farm. And they do it because they need a quality and character of life that farming PROMISES. Some might say it is not a genuine need but rather a suspected preference that motivates people to select farming. That is not the case. Medicine and religion have identified that people locked into meaningless lives get sick in body, mind and spirit. Some, not all, of those people have been fortunate enough to deliberately or accidently discover that their way of living has an enormous effect on their well being. Once that discovery is made, and additional information and experiences feed it, an urgency develops and what may have begun as a teasing suspicion becomes a clear 'NEED'. Given that humans thrive on promising challenge and farming holds the possibility for health and happiness the question goes from "why farm" to "why not farm".

Who Shall Farm

The last quarter century in this country has seen a parade of ugly propaganda that has confused the non-farm populace about this most noble of vocations. It would require volumes to enumerate and repudiate every piece of garbage that has slipped into or been forced on our society with regard to the culture of farming. But so much of it comes out in the notion that farmers, as a group, are dullards, slow, unsophisticated, backwards, with even the hint that they are inherently retarded. Sure, it is granted that they are honest and hard-working but it is added that "they don't know any better". What a disgrace that it should become axiomatic in our culture that anyone who works hard and is honest "doesn't know any better". And, actually if you think about it, what could be better than hard work and honesty? The problem lies in the fact that the connotation suggests a choice to farm is a step backwards. So our contemporary culture answers the question "why farm" with "why withdraw from mainstream life to a business that has no future and barely provides a living".

Throughout our history there have been sterling examples of individuals who deemed farming to be the most important part of their lives. And to this day the best, brightest, most courageous, healthiest, strongest and most honorable citizens reside on, or claim as their background, farms. The "Hee Haw - Yuk, Yuk - Hick image of farmers must be seen for what it is, a defense posture taken by those who feel their place in society is somehow threatened by farmers quietly displaying the best of ethics, morality and well-being. How else do you explain the urban posture these last few years that seems to say that farmers experiencing more than the usual dose of misfortune have it coming. While the same people embrace the few examples of brashness and vulgarity found in farmers ranks by providing space and time in the cheap and dirty media of TV and newspaper. (This subdued but smoldering mistrust between farm and non-farm cultures will be an invested thorn in the side of our society for years to come). Farming remains, even through these tough times, the most honorable calling requiring the best of each individual.

Standards of Living

It has been said that if you want to make a lot of money for heavens sake don't farm. It should be added that with intelligence and hard work farming can, and regularly does, provide an excellent standard of living with a good net income. But just what does 'standard of living' encompass? And how is 'net income' measured? If you are warm, comfortable, well clothed, well fed, working closely and harmoniously with your family, able to care for your health needs, enjoying a satisfying, interesting community social life and feeling good about your tomorrows - your standard of living is excellent. Farming can and does provide ample opportunity for all these things to happen. If at the end of a year your freezer, cellar and/or cupboard is well stocked, your barn is full, your house is warm, your bills are paid, seed stock has been saved back, anticipated expenses for the coming year are covered and you have cash 'left over', your net income is very good. That net income does, by all rights, include not only what cash you have 'left over' but also some measure of your gain be it in replacement heifers, seed oats, new fences, etc. An urban factory worker may make 15,000 dollars or more before taxes each year but after all bills are paid what kind of 'left over' cash or accumulated gain does he or she show? And to what extent does that city life 'create' necessities out of what farmers might consider luxuries. Money is a medium of exchange. To most people that translates into working for money to be able to buy groceries. If a farmer grows a portion (or all) of the needed food, he or she has made an exchange and money may have paid a negligible part. The primary medium of exchange, in this case, would be a combination of energies including human labor and creative energy. That is but one example of why cash income is not an honest measure of standard of living for farmers.

Since world war two, many good farmers have become confused about just how their standard of living differs from that of the non-farm labor force. And that confusion did not originate with the farm community. In many ways it can be said that it was the direct result of government, banking and industry working to homogenize and control agriculture, that last bastion of the independent human spirit at noble work. Officials of GBI (governmental, banking, industrial complex) have been, are and continue to be threatened by the notion that any measure of food production be left in the hands of millions of people viewing their work as a sacred trust and enjoying the life of it. They have worked to industrialize farming, even socialize it. They made all manner of alluring promises of greater income and security. Gov't. programs were enacted to provide "guarantees" and "help". Against rapidly appreciating land values, hundreds of millions of dollars in loans were made 'easy as pie', and farmers were told to get more land, newer equipment and out produce each other. In their rush to take advantage of all this 'good fortune' farmers by the thousands mortgaged their land, equipment and livestock to a staggering extent. And even worse, they mortgaged their future and their heritage for their piece of the empty promise. We all know how today's chapter is unfolding with thousands of foreclosures, liquidation auctions, divorces, suicides, alcoholics, and massive depression of the spirit and of small town America. It can all be traced back to those individual moments when the decision was made to borrow money and improve through accelerated growth. With those decisions came a denial of the standard of living inherent to farming. Farmers no longer had time to grow their own food, they were going after more money and could 'afford'(?) to buy all their food. They had a big net worth on the bank financial statement but their standard of living became progressively worse.

All of this calamity in today's agriculture is not without remedy or hope. If troubled farmers can somehow come to rediscover farming as an Art and way of life rather than as a subsidized industry the natural standard of living will fall back into place. There are, of course, many difficult individual circumstances which will require courage, stamina, and imagination in order that the farm family can either save the 'home place' or start anew. In many ways the problems of today should perhaps be viewed, and ultimately digested, as one would the barn burning down, or disease wiping out crops or livestock. It will be important to understand the cause of the destruction and work to see that it cannot happen again. The main point is that, although it won't be easy, there can be no doubt that it's possible and worth the effort.

Self-Worth and Contentment

As with any vocation, there are people in farming who do not enjoy the work, find the possibilities unexciting and long to be elsewhere. Perhaps farming will never be an attractive way of life for them. Yet for each one of them there are dozens who feel a born kinship for this wonderful and varied work. Unfortunately, not all of them will have the aptitude to be the cream of the farmer crop. Nonetheless, it's possible to live a farmer's life in true contentment with no chance of being measured by others as better or best. That's another special aspect of this business; it's an arena for the individual (or family) adventure allowing personality to determine the character of each day. And allowing that success be measured by individual goals and interests rather than production performance. People who enjoy the work of farming will find the rich rewards in every moment. Those who do not enjoy the work will find an endless stream of unpleasant experiences.

With those who have been troubled by recent calamity,

rewards may be difficult to see. Vision has been clouded by a sense of failure, or worse hopelessness. For them a difficult time of re-examination has come. It is critical, if farming is to be restored to them as a heartfelt and enjoyed trust, that they come naturally to see once again that their way of life is within their control. Any true and lasting sense of self-worth and contentment must come from within each of us. If it is triggered by external influences (such as a banker telling you you're doing a good job and offering a bigger operating loan) you have made yourself extremely vulnerable as your good feelings will be short lived and prelude to certain fall. Just as the health and success of an individual farm must be based on careful growth from within, so must the mental well-being of the farmer be based on affirmation that comes from within.

Fabric of Life

A life of farming does, if allowed, have a critically profound affect on the fabric of life for the farm family. Agriculture encompasses such an incredibly diverse spectrum of possible ventures that it is not reasonable to think we can touch on each. But touch we must because, however limited, it may tell something that all these abstract words that have come before have missed. These are some of the sorts of moments that can add to the weave:

As you look over your pen of market lambs a smile crosses your face because you have finally come to see your ideas and labor manifest in this fine set of uniform stock...

Your back hurts and your hands are rough as the last load of hay is put safely in the barn and it's hard not to enjoy the feeling that winter has once again been prepared for...

It dawns on you and an announcement is made to the family, all the food on the dinner table has come from your farm and garden - the children smile because they feel for you what that means...

A neighbor compliments you on the flavor of your produce as they make an order for several lugs of cucumbers from the next pick...

Your daughter comes racing into the kitchen one morning squealing with delight that the pullets have laid their first three eggs...

You stand near the gate and marvel as you watch the young stock dog carefully bunch the heavy ewes...

Milking chores are done and your thoughts are with the fresh strawberries topped with real whipped cream waiting in the kitchen...

Your son, without coaxing, hurries after school to change his clothes so that he may feed his pigs...

You stand with your neighbors and take a moment to look over the barn you have raised together...

The posts are set, the wire is stretched, and the gate is hung. A new field has added another dimension to the look and feel for the farm...

The team of work horses stand quietly and you fork yet another load of manure on the spreader thinking that Winter always promises a rest but seems always filled with postponed projects that can't possibly be completed before Spring...

You walk through the grain field calling on eyes, fingers and teeth to tell you when harvest will come...

The last pass is made over the field in preparation for seeding and you look over your shoulder with a recent mind full of each tedious yet pleasant procedure...

On this hot Summer day you and your entire family are fishing together and eating peaches from your own trees...

Your daughter can be seen out the window pointing out pieces of the farm to your visiting grandchildren and you realize that she was affected by her raising in ways you hadn't suspected...

The red-tailed hawk hovers with flashing wings, the setting sun moves through the Autumn clouds shooting rays across the corn stubble, a bell can be heard as the Jersey trots to reach the milking parlor first, your boy - stick in hand - waves to you from behind the last milk cow, one of the twenty that move in single file to the grain pot, the chickens start to move up the little gang plank and into the hen house for the night, a ray of the setting sun escapes, glances, quickly, across the tops of the pumpkins and into the faces of two sheep peering through the fence, the dark silhouettes of the equipment stand, - beside the back of the barn, - like metal skeletons forming a backdrop to the little battlefield that is the hog yard, you see your daughter and her boyfriend sitting close together on the back of the old hay wagon teasing each other about the future, you feel the hand of your spouse slip into yours and you look into a familiar smiling face, you're told that the brindle steer has just busted the new gate near the creek and that the neighbors will be coming by later to talk about the irrigation project, you look up and the hawk has disappeared, the sun has gone with only a reminder of its former light in the colored underbellies of the soft clouds and you head for the milking parlor together, in the parlor

window's light is the outline of your boy's face as he peers out at
you and you ask;

Why farm? why, because there can be no finer life.

Chapter Three

The Future of Small Farms

Will Small Farms prosper or suffer during the 1980's and beyond? The general social mood in North America is changing. And with that change of mood seems to come a change of values. The FAMILY unit is under increasing threat as a misunderstanding and misapplication of personal liberties chisels away at the basic footing of society. With this growing "me first" attitude comes, ironically, a growth of the socio-corporate-police state. For some those are hard words, for others they are thought to be ridiculous. They are used to characterize dangerous and idiotic developments and evolution within local, state and federal government as well as the corporate boardroom. In this environment of stagnant metamorphosis the Small Family Farm is an anachronism. It serves as a powerful, yet awkward, example for constructive progress in the future.

A different version of these thoughts originally appeared in Vol. 5, No. 3, 1980, of the Small Farmer's Journal.

But Small Farms will not survive nor prosper just because they are good models for society. They have to successfully compete within our industrialized-socialized society.

The psychological condition of our Western society is a subject that is as complex as it is forbidden. People do not want to hear that they are selfish or spoiled or insensitive or greedy. People do not want to hear that their mental state is dangerously close to being vacuous. Far too many people do not want to think about the consequences of their actions, about the human condition, about the health of the planet. This is part of a slow corrosion process nearly complete in our cities and coming on strong in many rural areas. It is in the cities that we can see clearly what the results of this new 'consciousness' will be. We see greater alienation, loneliness and new forms of mental illness.

This condition is a terrible fungus destroying our social fabric. Government and large multi-national corporations do not see it that way. We have become a very easy people to control (or govern if you prefer). You see, if we were all very hungry and poor we would be extremely difficult to govern. We would only have to witness the slow death of one of our children from starvation to go after the throat of a bureaucrat or wealthy corporate executive. But the mass of us are not hungry or poor. In fact, we have become collectively insensitive to the relative hunger of our own people. We ask too often, of our own poor, "why don't they help themselves?"

With a few notable exceptions, state and local government has lost any semblance of pattern or direction. As we forfeit our "collective" consciousness to the "me first" vacuum we see the development of stupid, ill-conceived, shortsighted local government which bounces from one special interest to another.

Federal government enjoys having to react or respond to sophisticated, well-oiled special interest groups who can deliver them the "reason", the "money" and the "vote". The bureaucrats and politicians can then give us the "reason" secure in knowing they have the "money" and the "vote". And government programs are easier to implement if there is no appreciable struggle with the masses, no dissent. And the nature of the programs and legislation is often subtle and insidious, as with the Seed Patenting legislation which passed last session.

SMALL FARMS must look to themselves for preservation as in this social and governmental climate we cannot expect any help. In fact, we can expect a new campaign to;

"separate the 'real' farmers from the 'phonies' and get those inefficient bums out of agriculture altogether".

I fully expect to see some new language develop so that the Feds and Multi-Nat. can identify the true 'small farms' in some negative way while referring to the medium sized family farm as 'small'.

I should say, after all this bleak stuff, that I feel the 'outlook' for the Small Farmer is extremely good. That takes into consideration the strength of character that appears to be a common denominator amongst small farmers. But we are going to have to work harder and smarter in order to keep going. We need to take long careful looks at successes within our own ranks. Of course we need also to identify failure and use that information with equal care. We need to DEMAND more of those public and private institutions which claim to serve our needs. And we need to be careful and always consider the source of information. The USDA, in particular, continues to provide information that seems to be deliberately inappropriate or even detrimental to small farm operation. Equipment and chemical companies will spend whatever money they need to in their constant struggle to get you as a dependent. We need to identify changes in conditions and how those changes affect us whether that be in regard to access to land and money, or openings in the market place. And marketing may just be the most important area of change to affect the small farmer in the '80's.

Whether you are conscious of the fact or not, one critically important aspect of your choice to be or stay a small farmer is that you are also by definition a public servant. Some will doubtless find these words offensive (I do) but it represents the difference between a "me first" mentality and a "what can I do to help?" sensibility. So small farmers may, in their continued survival, offer direction out of the Orwellian era.

Chapter Four

THE LIVING FARM

farm, n. piece of land used to raise crops or animals. -v. 1. raise crops or animals on a farm. 2. cultivate (land). 3. let for hire.

living, adj. 1. having life: being alive. 2. full of life: strong. 3. still in use. 4. true to life; vivid. 5. of life; for living in. 6. sufficient to live on. -n. 1. act or condition of one that lives. 2. manner of life. 3. livelihood.

Familiar words often mean different things to each of us. Take the word "farm", for instance. Depending on your viewpoint it could mean a "piece of land used to raise crops..." in the starkest, most absolute sense. Or it could mean "a fertile place with life of its own on which a wide variety of crops and livestock depend for home." Connotation, or what we see included or excluded from our mental picture of the "meaning" of a word, can often be more important in communication than the dictionary definition of the word. Also, common practice can often add to or take away from the meaning of a word. In this case we're interested in what has become of the word "farm" today.

There are enough changes going on in our world today to give at least two different "feelings" for the word "farm." "Agribusiness", a new word (if you can call it that) coined by economists to replace "farm" as a word, goes a long way towards suggesting meaning. The farm is viewed as a food and fiber factory where advanced technology (heavy machinery orchestrated in patterns of overkill) and chemistry (deadly poisons and unnatural growth stimulants) work with only a little help from soil and weather towards the most "efficient" (another word with changing meaning) mass production of raw materials for eventual use as food and fiber. Another dimension to this new meaning is that by sheer size (not health or true profitability) this sort of "farm" commands the respect and attention of state and federal government. This relationship serves the best interests of both agri-business and government in many important ways. And, it is assumed that "if it is good for 'agriculture' (you notice the word has been substituted) then it is good for the country". And if that in and of itself were not enough it is said "if it is good for the government, it is good for the country." So in this context, what is suggested in addition to the simple meaning, is quite awesome.

But there is another side to that picture of "agri-business" that needs to be touched on if only briefly. These farms are dead empty factories. They have close-ended production cycles. In other words, that only positive relationship that the last crop grown has to this crop growing is its effect after sale on profits. There is little or no maintained on-going inter-relationship between life-cycles on these farms. Because of this, soil condition is such that maintenance of maximum production levels requires annual increases in the input of energy and chemicals - raw materials, all from off-farm sources and purchased at wholesale/retail prices. The affect of this, and the ever growing need to replace obsolete and old (5 yr?) equipment with new, larger, faster, more all-encompassing machinery, is the requirement of large amounts of short term capital. In studies across the country it has been found that the measure of debt incurred for equipment, fuel, chemicals, seed, and such, often exceeds or equals the value of the land farmed. Not long ago good farmers considered such indebtedness unthinkable. All of this serves to reduce the margin of profit and require that the agri-businessman (he's no longer a farmer) expand his operation. This expansion, of course, costs money and so...

We've only touched on some of the business aspects of this "meaning" but perhaps it's enough to suggest our prejudiced view of that connotation. By contrast let's talk about the "other" meaning.

At the head of this article we included, after the dictionary definition of "farm", a definition of "living". The best way to give a full suggestion of what we want of the "other" meaning is to couple the two. The "living farm". Although we use "living" as an adjective there are aspects to its meanings as a noun which add fullness to the connotation we're after for "farm".

THE LIVING FARM

"piece of land-having life; being alive" Land used to raise
crops (and) livestock - full of life, strong.
living...still in use; true to life; vivid, of life, for living;
manner of life; livelihood

For the purposes of this discussion, we need an outline of
what it means to be a living farm. I said "outline" for good
reason. If we "define" it we limit it. The outline should better
allow us, mentally anyway, to include all that is appropriate. And
this inclusiveness is an important aspect of the sort of farm we're
looking for.

This outline must include all the dimensions of stock-
farming, and crop-farming. It must have room for the inter-
relationship of both. In fact that farm would be the interaction of
both. We are wanting a farm that will serve as a living piece of
self-renewing fertility.

Crops and farming are almost universally thought of
together. It's easy to associate apples, wheat, corn or soybeans
with farming. But enough of a change is taking place in contemporary agriculture, that some have difficulty equating livestock
with farming. That seems bizarre to me. But we have to remember that the diversity of operation that allows for livestock has not
been a popular aspect for agricultural engineers in the last thirty
years. Monocultural practices or the production of single crops,
whether that be wheat or tomatoes, has been a popular approach

sold in the last half century by agricultural "planners". One of the most important intrinsic results of this change has been the removal of fences from the farm landscape. On farms where rotational practices had used to include an ongoing concern for livestock we now too often see animals excluded from farming considerations. Concentrated mechanized feedlots seem to be the direction that "efficiency" engineering is pushing us.

The living farm is based on an elaborate inter-related family structure, including families of livestock, families of crops and the farmer's family all living together. The result is that the farm itself has an ongoing regenerative aspect that gives it a life of its own.

Allow me to give one abstract example without limiting the design of all living farms to this:

Imagine that our living farm has "X" number of acres which are rotated to grow some wheat, oats, barley, corn, pumpkins, beets, and hay. Also on this farm is a well established fruit orchard of medium size. The entire farm is fenced and cross-fenced into relatively small fields. All of the field work on this farm is done with horses. A stallion and mares are used and colts are raised for sale and replacement. Oats and hay are used to feed the horses; home grown feed. A small herd of dual purpose cattle are maintained with calves raised for sale and cows milked. The milk is sold off the farm in the form of cream or butter. All the feed for the cattle is farm-raised. A small flock of sheep is kept, with wool and lambs not used by the farmer's family sold. Hogs are kept on pasture when available, in the orchard after harvest to clean up rotting windfalls, and fed surplus milk and other crops in the winter. A flock of geese roam the fields keeping young weeds down. And poultry is kept for eggs and meat-there again, all feed is farm raised. The farmer's large family provided the labor and reaps not only whatever dollars may come from the sale of the various products, but more important the finest manner of life. They reap a "true" livelihood providing a quality of experience and environment that few other endeavors can match. The farm itself grows in fertility not only by the addition of animal manures to the topsoil or by the various positive effects of crop rotation but by virtue also of the fertility and health of the crops, livestock and the farmer's family. You see the farm has a life of its own.

Again let me say that the picture I paint is just an example, any number of various combinations of crops and livestock might be assembled with the same resultant fertility and health.

Another point to be made is that if you can break loose of the simple linear economics that we're all used to, (example; cost 'X' dollars to raise crop - sold for 'X' dollars, shows profit or loss when subtracted) and think in terms of a broader interrelated economics with concerns for "true" costs and "true" profits whether or not they carry a market (dollar) value, then the living farm we speak of has a much greater "viability" or "efficiency" than it might commonly be credited with.

If a farmer, by choice, is to control and/or orchestrate this farm we speak of with a concern for its ongoing life and health, he must become a special sort of maintenance man. A generous capacity for, and understanding of, "maintenance" as it is related to the farm in the broadest sense is perhaps the most critical characteristic that the farmer must hold. Not only must the farmer maintain buildings and fences but also he needs to concern himself with crops and livestock. Perhaps most important, however, is the ongoing maintenance of the soil and its fertility without which the farm would be baseless. And overriding the particular concerns of a given moment, the farmer needs to carry with him always a plan, an idea, an ideal, a concern for the ongoing health and life of the whole farm. The farmer needs to be a special sort of maintenance man. The farm must be alive.

Chapter Five

Knowing How To Work

In April my six year old son, Ian, went out to the ranch with me on a quick trip to pick up some lumber. I've got a "hen-house" sawmill outfit on which I had cut some cedar barn boards for a friend. I had to load up a trailer with those boards. Young Ian wanted to help "Old Dad" so I "humored" him. I was in for a wonderful surprise and some come-uppance. I picked up a board, carrying all the weight and let my son get a "pretend" hold on one end of it as I set it on the trailer. He frowned and said,

"Come on Daddy, I want to REALLY help!"

I smiled and asked him if he thought he could pick up one end of the board himself. He grimaced and grabbed hold of it. I was surprised. With him on one end and me on the other, we loaded those long boards on the trailer. He REALLY was a help! A big help. Sure it was hard work and he got a couple of splinters, but he was enjoying it. At one point he stopped, took a deep breath and said,

"You know Daddy, I'm not smart, but I do know how to work."

I'll never forget that.

Later, as we were making the long drive back to his mother's house, he leaned up against my side and said,

"Daddy, how come we can't spend years and years together working on the ranch? You know I like hard work with you."

I told him that I'd like nothing better, but that I had lots of bills to pay and a lot of a different sort of hard work at the Journal. I told him my goal was to get the bills paid, the Journal doing better so that he, his brother and sister and I could spend years and years together doing "hard" work on the ranch.

Two weeks later I got a letter at the office from Ian. In big pencilled letters, the note said:

"Dear Dad, I love you Dad. I know you are working hard. I love you Dad. Love, Ian."

Taped to the bottom of the note was twenty-five cents, his allowance. He sent me his allowance.

It choked me up.

Chapter Six

How Do We Rescue The Small Farm?

T he much heralded news of the demise of the small farm, the death of the family farm, the obsolescence of all but large-scale agri-business, is an organized lie. The news comes to us in television commercials, county extension agent reports, USDA reports, chemical company reports and advertise-ments, from food-producing and marketing conglomerates, from so-called experts in agricultural engineering and on and on. In every case the source of the news is an exponent of large-scale agri-business with a sizeable vested interest in its continued growth. This, coupled with the fact that in almost every instance the news does not talk of the eventual or inevitable, but almost as if an aside, they mention continually the "recent" death, the "demise", the "current" obsolescence.

Anyone with a working familiarity with the small farmer population in this country would be hard-pressed to agree with

Originally appeared in the Spring 1977 Small Farmer's Journal

this assessment. The small farmer is alive and well and numbers in the millions in this country. We received recent news that the USDA and IRS have raised standards for what they consider a legitimate farm. So are we to believe that because the government doesn't count us, we don't exist?

We must ask why, whether it is true or not, is it so important (important to invest many millions of advertising dollars) to mount such a massive campaign on many fronts, constantly reminding Americans of what "has happened". Since it is not true, and the recent continued growth of rural populations makes it a larger and larger calculated lie, it must be assumed that something is to be gained by this propaganda program. A program funded by corporate interests and government agencies. And on the government front, it is a program that has not changed its tenor for decades during both Democratic and Republican administrations, indicating that it is a policy which overrides political ideologies. Where the corporate concern seems to be limited to the advocacy of large-scale agri-business as opposed to the small independent farmer, the federal government policy continues from that point on to include the belief that it is justified in manipulating production and marketing because of the new-found leverage that controlled agricultural exports affords it on the international scene.

In both cases, this policy represents a very serious threat not only to the small farmer, but more important, to all rural communities, regions, states, and the country.

If it were to succeed, it is envisioned that agriculture

would become a network of mammoth units taking advantage of computer and aerial technology with centralized headquarters utilizing remote control. Plans are on the drawing board to incorporate nuclear technology. Livestock would be raised in factory-type housing, never seeing pasture. Fields would be huge. Fences, houses, barns and trees would be leveled. Massive crop storage facilities would be centralized. Marketing would be handled through brokerage firms in the cities with the Futures markets playing a very large part.

The result of such a dream? Rural communities would lose population, customers, regional influence. In other words, towns would die. Farmers would move to the metropolitan areas unemployed. Consumer options for food sources would be limited to retail super-market outlets. Controlled ownership of land would make the small private land-owner near obsolete, certainly in the most productive valleys of the country. And the quality of food-stuffs would be determined by the corporate ethic. And that's not all, the results would be ghastly all across the country, affecting eventually everybody. It is not a pretty picture.

Thank heavens, it hasn't happened yet. And it can be prevented. We have some important conditions in our favor if we use them properly. The most important factor on our side; the energy crunch.

But first a look at the most dangerous aspect of the threat. Sometimes it seems corporate interest and the federal government are working together by accident. But, far and away, the most dangerous threat would come from the government if it should choose to step in with controls and tax advantages in favor of large holdings and corporate management. During election campaigns it is not uncommon to hear a great deal of mouthing praising the contribution of the family farm. Yet in Washington, D.C. the legislators and executive branch praise and reward the huge corporations, the conglomerates, the multi-nationals. The executive branch even patterns itself after corporate leadership organization and the newly appointed cabinet has more than a few former corporate heads in its make-up. Certainly the government in its deeds has been a strong exponent of the corporate ethic. An ethic which puts the success and the "integrity" of the company above all else.

THE SMALL FARM IS VIABLE

Well, I said this doomsday prophecy could be prevented, here is what I feel we must do. Most important of all, we must believe that the small family farm has worked, does work, and will work. We must believe in the positive effect our success has on our community and the country. I know that sometimes those effects are hard to see, but they are there as long as we are. The reason that this belief in ourselves is so important is that it is, after all, what keeps us going. And we must continue.

Next and very nearly as important; Farmers and farmers-to-be must do some predicting of the shape of the future, both near and far. And we must make some careful decisions based on those predictions. Farming and stock raising have always been long-range considerations. Without bags of money to buy additions or changes, farmers must plan those changes into their long-range scheduling. With most livestock or crop, additions or changes, a minimum of three years must be considered if you choose to grow into them. It takes time to raise replacement heifers for a cattle herd and it takes time to pay for a new barn or establish an acre of strawberries. So it is important that each farmer plan his future and in those plans take into careful consideration the increasing pressures applied to drive him out of business. He must plan for his survival. It is imperative to consider all the available options that might aid by providing greater flexibility, independence, and general security. We must be prepared to make changes and adapt before we're forced to.

Three areas will make us increasingly vulnerable in the future; TAXES, MARKETING, FUEL/EQUIPMENT.

TAXES. As the saying goes, taxes are inevitable. With the small farmer who owns or is buying his land, the taxes on that land often represent a largest and critical on-going, out-of-pocket expense. And that tax, more often than not, is on the steady increase. As a recognized problem, taxation is the most difficult one to have a constructive effect on.

The small farmer must understand all the particulars of his tax structure and take an active part in those elective and public hearing processes that determine government spending and tax rates. This is the only way he can enjoy a measure of effect on his taxes.

MARKETING. The economic success of even the best

of farmers will always come down to the marketing question. Under the current system, or non-system, in this country, if the farmer has good weather and a successful crop, surplus drives prices down. When the weather is bad and crops fail, prices go up.

That, of course, is a generalization, but, for those farmers large or small, who are dependent on "established" marketing channels, the pressures are very real.

In order for the small farmer to insure a greater measure of his own security, he must break away. These marketing systems are established by, maintained by, and exist for agri-business concerns and will continue as such. There are a few exceptions in the form of regional farmers' cooperatives, but they in turn are dependent on the next marketing level.

The small diversified farming operation has the distinct advantage of having several commodities to market, thereby limiting vulnerability to supply fluctuations and erratic prices. And these farms must work to make their products available more often for sale direct to the consumer. Only in this way can they have any hope of gaining the added security that would come of marketing independence. Establishing new, more direct, markets takes time and effort, but the result should mean a slightly higher, more stable price to a regular clientele.

Some marketing ideas might include; direct on-farm sales to individual consumers and marketing clubs; roadside stands; sales direct to small independent retailers, such as greengrocers or butchers; small scale home processing and distribution.

Utilizing canning and drying processes and including mail order distribution; direct sales arrangements with local schools, convalescent home, hospitals, and so on. There are many more ideas and they should all be explored. The important thing to keep in mind is the goal sought by more direct marketing, better price and sales security and independence from established marketing systems.

In this country the small farmer has the undeserved reputation of being poor and inefficient. This attitude towards us is perhaps a result of our rapidly changing social structure in this country. In western Europe the reputation of the farmer is quite the opposite. And there, in France, Germany, Switzerland, Holland, to name a few, the "recognized and appreciated" backbone of agriculture is the small farmer. Those farm families have built-in security and self-sufficiency to such a fine point that for generations they have survived the calamities of weather, war, and society, very often on small plots of ground that have been in the family name for hundreds of years. For example, some have said that these same characteristics were responsible to a large measure for France's ability to withstand her occupation by the German armies. Hitler seized the cities and thought he had France by the nose, as it turned out, he had her only by her ornamental necklace.

And these European farmers sell their goods direct more often that not. But that is not the sole reason for their health. For whatever reason, whether deliberate or not, they have not mortgaged their land wholesale to over-equip it with machinery and technology. Which brings us to the third concern.

FUEL/EQUIPMENT. Farming today, large and small, is critically dependent on petroleum products and, to a lesser degree, natural gas and electricity. Without readily available sources of these at reasonable prices, agriculture in North America is helpless. Should so called "temporary shortages" return, it is well within reason to suspect that the federal government would implement rationing. Accepting as the government position, a belief that greater size is followed by greater efficiency, and that promoting so-called "efficiency" is important to the country, it is expected that any petroleum rationing would favor agri-business over the smaller family farms. Though we may not accept this "efficiency-follows-size" logic, we may still find ourselves forced to accept a very real shortage of fuel. Of course, any prolonged severe shortage would affect everyone in this country.

With a fuel shortage, farm machinery prices would again rise. As they stand now, most new farm equipment prices are so high that the farmer cannot hope to pay for it through farm production. And it seems that the American farmer, large and small, has a dangerous love affair with tractors and assorted machinery. Dangerous because it makes economic security an uncertainty by narrowing both long and short term profits. The bigger tractor takes more fuel dollars, more maintenance dollars, and the farmer has got to lease or buy additional acreage in order to pay these increases, plus the payments on the tractor. And the greater the acreage and larger the equipment, the smaller the profit per acre. It is a nasty cycle which few dare to break. It takes a courageous, sensitive, intelligent farmer to limit the size of his farm to what he knows is appropriate and right. We're not arguing against growth, we're arguing for new definitions of "efficiency and productivity".

When the questions of size, fuel and equipment enter the picture, it is time to discuss briefly horse-farming. In this country the mass of people consider farming with horses as either a great joke or a sweet bit of nostalgia. But for the small farmer, concerned about his critical dependence on fuel and machinery, horse-farming offers a practical, viable and exciting option.

We'll just mention here in passing that with a horse operation a good farmer can look forward to less overhead cost, no long-term equipment debt, nearly total independence from petroleum products, better soil condition, and additional livestock for sale. In short, the horse-farmer can have a healthier farm and keep more of what he makes without ever "having to grow big" for survival's sake.

For those who don't choose horse-farming, some suggestions for limiting fuel/equipment vulnerability might include; setting up fuel storage systems on the farm and keeping them filled to cover short-term shortages; selecting used tractors that provide the greatest fuel efficiency; using smaller tractors; limiting tractor operation to necessary field work; combining field operations such as discing and harrowing or seeding and rolling; maintaining a well equipped farm shop for repairs; buying only used equipment and only that which can be paid for; planning a cropping system with a concern for limiting tractor use; and much more. In these ways the tractor-farmer could greatly reduce his fuel consumption and his future needs.

These three areas, Taxes, Marketing, Fuel/Equipment are the most sensitive for every farmer, large or small. It's often heard that the one thing that makes farming such a risky business is the high cost of land. I don't agree wholesale, not when the combined cost of taxes, marketing, fuel/equipment over a period of 15 or 20 years far and away exceeds the value or cost of the farm acreage. (Certainly in some suburban areas of the states the cost of land is prohibitive if it is to be farmed.)

But how do we measure land against a 180 horse-power four-wheel-drive tractor with eight tires and an air-conditioned cab priced at $100,000 and carrying a life-expectancy of seven years and a depreciation scale of value which is straight down.

The land, the soil, must be returned to its proper place of all importance in agriculture. It seems wonderfully ironic that the methods and changes necessary for doing this would at the same time grant the farmer greater security and independence.

SOME CLOSING THOUGHTS

Historical statistics indicate that during the first half of this century a large migration of farmers from every size of operation and every corner of this country went to urban and suburban communities. In fact, the migration was so large in the first thirty years of the 20th century that some historians and economists believe that it caused a terrible social and economic imbalance which made an important contribution to the Great Depression.

It is an interesting side-note that quite a few conscientious and intelligent farmers of the '30s predicted that the wholesale replacement of work horses and hired men with tractors would cause massive long-term unemployment and huge farm commodity surpluses, the short-term result of which would be a depression. The long-term result would be the centralization of the country's population, causing continued social and economic problems. Problems that some have prophesied would take a century to repair and could well cause wholesale socialism.

The last ten years have seen a tremendous migration back to the farms and back to rural communities. People are seeking a better life and seem to believe it's in the country. However, we've seen an odd turn of events with this migration. These people seem to have resigned themselves to a particular economic formula. The majority speak of their inability to make enough income from the small acreages they own or lease, but this doesn't seem to stop them. They accept it and go from there, securing outside jobs to bring in additional monies. This peculiar pattern seems to complete itself with these people longing for what they consider the dream position of being able to sustain themselves and make a living as farmers without the outside employment which robs them of their time. Whether it is in Mississippi, Vermont, New Mexico, or Montana, all across the country this pattern repeats itself.

It's disturbing. On the one hand we have this tremendous resource in an eager group of new farmers already on the land and on the other hand we have this apparent self-denial of ability. Now, no doubt there will be many who will take offense at this. Please accept it as constructive criticism. Those in this situation of which I speak must take the next step. And that is to look hard and long at what is there in terms of land, potential crops and livestock and the marketing questions. It is not good enough to limit oneself to that abstract outline for self-sufficiency which only takes into consideration the production of food for the family; the "new homestead formula". You must expand within the limitations of available land and resources to the production of farm commodities for sale or trade.

We feel the greatest tangible restriction on these people, young and old, to this end is their inexperience and the difficulty of finding practical, comprehensive information on small-scale farming (rather than "homesteading"). But it is a deficiency and a difficulty that must be surmounted. Those of us who know and believe in farming must reach out and give answers. It is our obligation if we do believe in the future of the small family farm and wish to help in its development.

Chapter Seven

From The Editor's Haymow

As I sit and write this, it's a hot night in late July. The children are in bed. It's quiet. The horses, cattle and sheep are enjoying what coolness the dark has brought as they wander in search of special discovered meals. By now the Elk herd has quietly left the shelter of the wood to join the livestock. They will be walking their curious camel-like walk listening for danger. The dogs are asleep and ten miles west the ocean fog will begin its slow, cooling crawl over the low mountains until it spills silently into our little valley.

Maybe down by the barn a porcupine will be lumbering across the road. Mountain boomers have left their hobbit-like den holes and are wandering the forest floor, under the security blanket of night. The children are asleep dreaming about watching the giant wads of loose hay moving up and into the big mow of the barn. They hear the grapple forks clank free as I jerk on the trip rope. And now they're thinking about tomorrow's watermelon and the bike ride to Wendy's house. It is the season

Originally appeared in the Fall 1984 Small Farmer's Journal

of the sweat, the rash, the heat you can taste, and it is the season of preparations.

My daughter, Juliet, lay in a pretty, crumpled little pile on the big chair, and wiped her forehead saying, "Leave them alone Daddy, I'll get those dishes in the morning."

I looked at her and felt a smile take over. A picture from Spring flashed thru my mind. Little Juliet was holding a lamb and asking if she could keep it for herself, and I thought, "may I keep this moment for myself." And again, as I noticed her tired in the chair, I wanted to keep the moment.

Those feelings must share time with this season of preparations. That's what haying is all about, preparing the farm for the season of maintenance. Some see four seasons. I feel two.

My thoughts wander ahead and wonder back. Ahead of me is excitement. After fifteen years, I still enjoy a young enthusiasm for planning and anticipating what might be accomplished with the farm. And behind me I can only wonder at the difficulties, confusion and anguish. I wonder at the strength of the draw that farming has on me. It has survived countless reasons to quit. I wonder at the friendships that have made it possible for me to continue, and I enjoy the prospect of repaying every helpful kindness.

These fast yet long summer days are filled with hard work and worry, and they are frosted with real happiness. My kids are with me on the farm, a new family - the Wagners - are sharing the work load. There is field work to be done, meals and laundry to be muddled through and, yes, a magazine to put out. There are bills to be paid, and I'm broke (aren't most of us?) Our weather is smiling on us, the stock are sleek and fat and things are getting done. My children smile and laugh a lot these days and I've gotten thru the first soreness to a fitness that surprises my age. Markets are down, so my new hand, Dale Wagner, and I sharpen pencils and work on ways to be better farmers and stockmen. There are fences to be built, barns to raise, fields to plant, logs to skid and lumber to mill. A season of preparations.

My neighbors can be seen every day in the hay field, or with the sheep or walking the Jersey to the barn. There is something comforting about having a good farm neighbor working as you work. Last week Dale and I cowboyed some cattle into my trailer for neighbor Haskell, and hauled them to market for him. He's been helping us with the hay. That was a good trade in spite of the fact that a certain blankety-blank bull worked hard at trying to make chickens out of us (I confess, he succeeded with me).

The beavers usually make a mess around here, but this time they've made a convenient dam on Little Soup Creek that's held together a nice pool near the barn. The wild blackberries are ripening fast which means the bears will be coming down for their August bellyaches. Yesterday, we saw a doe and fawn again in the hayfield, and Dale says the fawn has doubled in size since we last saw it. Down at Loon Lake, the water's warmed up and the large-mouth bass are stalking the shallows. There's activity at the Osprey nest high atop a dead snag overlooking the lake. My old barn owl has found a new summer home what with the hay track seeing so much activity. And tomorrow, my three kids will sneak over to the barn to leap into the loose hay giggling. Yep it's one of those precious seasons when you know, with each moment, you are building a memory that will sustain.

Chapter Eight

Time, Process, and Reason

Organizing Time

If you are doing what you enjoy, and most small farmers are,
the time seems to fly by. It gets much harder to get to every
project, every need, there's so much to do and each day
more to look forward to. That's a glorious state of mind to be in
when compared to the deadly, depressing agony of being dissatisfied and bored with your life's work. Every minute goes by so
slow and bitter when you're unhappy, lonely, dissatisfied,
disoriented. Most of us have experienced such a state of mind in
our own life. A few have yet to. Many are there most of the time.

One of farming's rewards is that it offers up to us a
beautiful way of life, a total day to day communion with the
creative processes of life. And we feel so easily a part of it all.
But then comes the pressure of time and it threatens to sour the

experience.

There are the new buildings to put up, the old ones to fix and paint. The soil to test. The drainage ditches to clean. The feeders to build. The equipment to restore. The fruit trees to plant. The poultry program to design. The watering troughs to set up. The pond to build. The limestone to spread. The pastures to clip and harrow. The sheep to cull. The mare to breed. The dog to train. The field to plow. The stumps to pull. The house to fix, plus more, more, more! And then the realization that there's just not enough time and slowly a dangerous frustration sets in. And that frustration threatens to undermine the pleasure of the way of life. It doesn't have to. We can outsmart ourselves in our pending anxiety. There are several important things to keep in mind.

First, remember what we are. SMALL FARMERS! We are that because we choose to be, because we can do a better, more conscientious job handling less rather than more. But ironically in order to do that we must do "more" with our "less". We must constantly examine the obvious, and hidden, potential interrelationships of every aspect of our farms. And the first question to ask yourself; "is my farm too big?" Almost before you answer yourself you need to ask the second question, "am I allowing all the parts of this farm to do their fullest?" In other words are you taking advantage of some of the built-in short cuts you may have on your farm? Are you allowing these short cuts to happen? Are you making work for yourself in the near future because of a lack of planning now? I know I do all the time. Here are some examples of ways of saving time and improving a farm's interrelationships.

Shortcuts

If you had to be away from your farm for a day or so, what are those essential chores that must be done? Make a list and include the average time requirement for each chore. This information is going to differ drastically from farm to farm. On a dairy there are the milking, feeding and cleaning chores which are constant year around. On a wheat ranch it's a different story. Most farming *Small Farmer's Journal* readers have mixed operations and chore requirements change dramatically from season to season. But the point of this exercise is to look at where we spend our primary chore time (and how much) and then ask if our secondary time is spent patching things up or carefully planning reconstruction or addition which will make everything run smoother. For instance, on a mixed crop and livestock operation the right layout of the fences and the best design and location of the barns and feeders could conceivably save hundreds of chore hours per year (30 minutes per day saved equals 182-1/2 hours per year); plus improve pasture and feed efficiency, plus aid in weed control, plus improve and ease livestock management, plus reduce internal parasites in livestock, all while making diversity easier.

The layout of fields and placement of fences and out-buildings should be done from a master plan. For instance, without the master plan, you might go ahead and plant a new orchard on the best and most logical ground, put a fence around it, and find afterwards that you have trouble using your pasture (just the other side of the orchard) for the dairy cows 'cause they have to go the long way around to the milking parlour. However, with a master plan you might find that the same ground with a different overall fencing plan would be easy and perhaps even advantageous to use for both orchard and cows. If you were able to save time each day because it was quicker and easier to move your cows, is that a convenience worth the investment in planning? The answer is yes.

Why not locate a pond just below and adjoining your garden with the orchard above the garden. Raise fish (and maybe ducks) in the pond and pump irrigation water to the top of the hill and into ditches which would irrigate the orchard and the garden. Set up the whole thing with terraces and well-constructed ditches to avoid erosion. You could go one step farther and have a windmill, top of the hill, above the orchard, pumping water from the pond into a stock watering tank. As the water overflows it is piped into ditches, irrigating as it goes downhill.

On any mixed crop and livestock farm fences play a big part. If you use fences to their fullest advantage, you'll have an assortment of small fields, probably of odd size, and likely you will have fenced lanes to comfortably move animals from area to area. Those fenced lanes should be considered and treated as pasture and the best tool for keeping grass and weeds under control in those lanes are sheep. With a little planning, a small flock of sheep can be utilized to save many hours of precious chore and maintenance time.

When planning time or making decisions about which jobs to do first and how to do them, always think things through to how your work will later benefit or affect the daily functioning of your farm. You will then be better off to make decisions about what happens first. The attitude should be less of organizing time and more of planning work and its results.

The Process of Farming

Talking about organizing time on the farm causes me to think about the process of farming; what it means, what it is.

Farming is a culturing process and more. Farming is kind of like the performance of an orchestra. The farmer is the conductor, the orchestrator. Using a musical score or formula as an aid, he or she must plant the seed or coordinate the conception while smoothly, fluidly, nurturing this potential life, guiding it carefully through an ever-growing maze of interrelationships. If it all works, the interrelationships serve up tonal harmonies and rhythms and patterns that make a whole of perfect, if misleading, simplicity. Though the musical score (or formula for crops and livestock) might be identifiably standard, each performance or season is unique in that it is a living process of creation. And the farmer is more or less artist as he or she understands or respects the involvement in the process as an opportunity to communicate vision while being lost in the urgency of getting it all done.

Farming is the most fascinating human endeavor imaginable but only if the senses are open to it. It includes everything that the human is capable of being aware of or confused with.

And in farming we are led to believe that we are either in charge or soon to be. But such is never truly the case. So it is an ultimate tease. Farming is almost a culturing process that certainly, and with certainty, gives life.

Chapter Nine

FRAGILE HOPE

On May 28, 1985, Ian Lewis Miller, aged 7, was killed in an accident on a neighboring farm.

Hope is everything to me. It sustains me. Knowing that the measure of my efforts weighed against the circum stances of my life gives proof of the body of my hope. I have always believed, until recently, that this hope was invincible. Many times when it seemed that odds were against me - and working, struggling and fighting were useless - I knew from within that I had to continue because there was always hope. This was never something I reasoned. It was a part of me, like my thumb.

With the passing of my son, Ian, I lost sight of that hope.

He was a delightful enigma. Tough, gritty, unafraid, sentimental and gentle. Many is the time I sat thinking or writing and felt his soft little strong arms encircle my neck from behind as he whispered into my ear; "I love you, Daddeo." I can feel it now as I write. Yet this is the same little man who would greet

Originally written in July of 1985

new acquaintances with an exaggerated grimace and a growl. He was the youngest and odd-man-out with his brother and sister. He handled that sometimes with tantrums and belligerence but there were the moments he won their pesky hearts by offering of himself. On one occasion, when he was left out of a game, he wrote a note to his sister, Juliet, which said, "I love you. If you tell anyone I will hide forever."

During the time I taught school and as a parent I took notice of how younger children find it unnecessary to stay with something whether it is a project, a game, or a plan. Ian always surprised me in how steadfast he was in his interests. From the time he was four being a farmer meant everything to him. At first I didn't take it serious. I was amused but tried to be cautious not to encourage him for fear it would turn into a damaging push. He paid me no heed. The farm was everything to him. And as I wrote in this magazine once before, working - honest to goodness hard work - with me just seemed to swell him up with good feelings.

He was self-sustaining. He'd go off to play by himself, entertaining as though he were two or three kids and instantly worrying me about his absence. Shortly before his death he'd come up with a funny little language of his own. He'd make a Bugs Bunny face and noises with complete phrasing. The day before he died he insisted on helping me clean sheep manure out of the barn. He was talking to me in his own language and I was smiling a confused grin. He said, "Daddeo, you can talk this way too. Just put your teeth over your lip and go "gedofleebo deedoodeedoo". I tried it and he said, "Yeah." with a big grin. When we were done I leaned on the pitchfork gazing at the clean floor and said "bedeepo flibit". Without looking up he said, "Yeh, it does look pretty good." And then he added, "Was I good help, Dad?" I miss him so much.

Two things happened to me the minute I was told he was dead. One of them nearly killed me, the other worked to keep me alive. With that moment a blackness enveloped me and destroyed my hope. I was a dead man inside.

The other thing which happened came of the necessities of the moment. There were things that had to be done, arrangements to make, people to contact. Friends and relatives wanted to do these things for me but some instinct drove me to action. From that instant I buried myself in work. Had I not, I believe I would not have survived.

In what now seems like a flurry of activity family, friends and hundreds of neighbors set to work to carve a little family cemetery out of a brush-covered knoll on the farm. We dug the grave and with special kind assistance from the mortuary buried Ian Lewis Miller on the farm he loved so much.

Almost immediately after that I set to work sawing lumber for Wayne Burch's barn. And then, in four furious days we put up the structure with a crew of five. After that it was back to my hen-house mill and more lumber sawing. Then there was haying and miscellaneous farm work. I just worked as hard and long as I could.

All through this time if I stopped to ponder, for even a

moment, business concerns - about getting a Journal out or answering letters or any of that stuff - I would cave in from the middle. Hard physical work, work that demanded my attention, was the only way I could hold my soul together. Two months passed this way, in a semi-conscious stupor. I had taken an emotional leave of absence. I was incapable of planning or abstract thought because it would open the door for the sickening terror to return. I was living without hope. Nothing sustained me but the raw inertia of hard work on simple projects.

During this time I discovered the immeasurable value of friendship. My loved ones were so care full and so careful for and with me. Justin, Juliet and Kristi carrying heavy hearts of their own watched and worried after me. Wayne and Joann and Britt and Robert and Leslie and my mother along with hundreds more just gave and gave and gave of themselves. And then there was my father, Ralph, what a giant! These gifts are beyond thanks.

I don't know when it started but at some point I began to test my wound and slowly, gradually forced myself to face the realities of my business world - this magazine. At the same time I looked hard at what Ian's death meant.

My life with my three children had been a cornerstone of my hope and my existence. We talk a lot, the children and I. We made, and still make, plans that center on this little farm of ours. Ian, Juliet and Justin were excited about my offer to help them start their own Miller Kids Chicken Company. They talked about it constantly. Without argument it was decided that Justin, the oldest, would handle the egg receipts and keep the books. Juliet would collect and package the eggs and Ian would do the feeding and watering. They would all share cleaning chores. Their enthusiasm for this project was total and seemed evenly spread. After Ian's death Justin, who is ten, said "Dad, I just don't think about the Chicken Company anymore." He does though. What he meant was that he needed time, I know that now. When he first said it, I took it to be yet another little sign that hope was dead.

As time passed and mending seemed to occur, I came to suspect that hope had not died with my son, not completely anyway. It had been badly broken and lay hidden beneath a dark shroud but it was still there. And something in me wanted to prevent his passing from being the beginning of the end. He deserved more than to have his life mean the death of hope for those who he loved so much.

It sounds like a maudlin cliche but I asked myself what he would want. The answers came easy. Above all else he would want us to stay with this farm and make all our dreams come true. He would want me not to worry or be so sad. And he would want us to see with each passing day how fortunate we are with all that surrounds us and makes up our world. With that realization came a flood of feelings which were headed by the gripping thought that my life would forever be the better because of my seven years and six and a half months with Ian. Then a light seemed to shine again on my hope. Just enough light to see its outline and the break. With that came a steadily growing determination that I would be all the best that was within me and more. Now as we say our silent grace before supper I picture Ian and see hope and give thanks. The shroud lifted.

One week after his death another passing hit me. The children and I raised, and were raised by, a Jersey cow we called Ginger. She was a real peach, a family pet. Some of you will wonder about anyone who could feel a genuine fondness for a cow, others will know what I mean. I discovered her, seven years old, dead in the pasture with no sign of cause.

Burying her was a trial.

So close after Ian's passing, her loss was lost in the larger terror.

On August 8th our Belgian stallion, Melodist DuMarais died a grueling death from a yet unidentified poison. He was seven and a family pet. He left behind a pasture full of his babies, broken hearts and wonderful memories.

One of Mel's colts, "Abe", is a two year old stud colt with the kind and gentle spirit of his daddy. When Ian was alive he used to play a game with me. He'd often ask "Daddeo, can I have Abe?" I'd answer with a smile "Son, he belongs to all of us." Only after Ian's passing did I realize that my son wasn't playing a game. He wanted Abe for his very own. He wanted to feel what that meant. I had unwittingly denied him that. I know what it meant to have Ian for my very own. So I've made Abe Ian's horse and I will care for him as long as he lives. He will stay here on the farm with Ian and the rest of us.

I have discovered in all this that hope is a fragile and precious commodity. It must be guarded, nurtured and honored. Those of you who read this; do not deny yourselves the strength that comes from knowing and sharing love you have with those you hold dear. Love them full, it will carry you through.

I dedicate my hope to Ian Lewis Miller.

Chapter Ten

Building a Radiant Culture

Mainstream industrialized agriculture is bankrupt in every sense of the word; it is immoral, socially and biologically irresponsible, and it is fostering cultural suicide.

Some would have us believe that 'great' civilizations eventually die because they are unable to withstand the infection that results from cultural identity crisis. I agree, in part, but take exception with the notion that the disease is absolutely terminal.

Those responsible for the state of our mainstream agriculture are not "Big Government (read Bureaucracy)" and/or

Originally appeared in Small Farmer's Journal, Vol. 6, No. 1, 1981

"Big Business" as we would all prefer to think. If you trace the whole mess down all the tangled lines you pass right into the scientific community, the labor community and the religious communities. But it doesn't end there, it goes beyond. Those responsible for the agriculture we have are you and me IF we support it by purchase and allow it by silence.

Cultural identity crisis is more serious and severe a condition these days than in any other period of man's history. The reasons, of course, have to do in part with media saturation. At their best and worst, newspaper, film, television, radio, popular music, books, and of course, magazines, might be mirrors of our culture. And, insofar as these media have no decent sense of proportion, perimeter or territory - no respect of myth, magic or religion - and are incapable of resisting profit - our multi-faceted cosmic culture is filtered through a magnifying mirror resulting in a vulgar confusing vision.

Big Government and Big Business are more vulnerable to public concern than has ever been generally believed. Look to recent events to see how fearful and quick our elected officials seem in the face of threats from home. And Big Business is scurrying these days to understand the changes in the marketplace. Our national economic condition, which some would believe is responsible for many current social problems, is not a mysterious social calamity. Our economic "crisis" is the direct result of over-reaction to misunderstood public outcry. And that 'over-reaction' comes because Politicians are afraid for their jobs and Big Business is afraid for its sales. Basic. Or simplistic? Disgusting. The makings of anarchy? What we need is to construct and accept a new precept ... or a new echelon.

There is contradiction here. If I am saying that we, as a culture, are getting what we "want" or "allow" - I am also saying that just as with a child in a carnival hall of mirrors, we, as a culture, don't know who we are or (most important) what we want. "Who we are" is a realization that cannot be sought. "What we want" can be, should be, understood and sought. (As what we want is sought, who we are will be understood.) But there is a problem of scale which defies collective understanding. If the boundaries, or perimeters of our culture were state lines (the culture of Minnesota, Texas or Maine) or even mini-regional (the Mississippi delta, the Oregon desert, the Dakota badlands), instead of the galaxy as perceived by CBS television or even the universe as mirrored by Twentieth Century Fox - the net results might include a decline in the sales of People magazine, the crumbling of vertical integration in agricultural industry, the slow death of Safeway grocery stores and hopefully the repeopling of farms.

If we 'want' to protect what is our civilization on this continent from the rotting disease of cultural identity crisis we will have to adopt a change of attitude or outlook. The United States of America may be some sort of political necessity but it should not be confused any longer as a cultural entity. Holland and Spain are part of Europe as both a continental and a political reality, but their cultures are distinct and different from each other and they are two of many cultures which are found in Europe. Like water and oil, these many cultures, complete with prerogatives, do not mix well. They stand apart, seemingly confident, each, in knowing who they are. So with this country, we must back up a little and take stock. How different are Maine and Arizona? Is that difference of value? Do we want, as a country, to be a nomadic state of overpaid voyeurs all in a constant state of flux moving from one place to another? Or are we ready to identify with a piece of geography, neighbors and a heritable horizon? Are we ready to accept the difficult discipline of culture? I think that there is evidence that most of us do want to 'settle down' and 'build'. But we must be careful to understand, somehow together, what we want.

If we can limit ourselves to an easily appreciated culture - one which does not extend beyond what we can experience as we live and work - one which radiates from the home and slowly flows outward only so far as the eye can feel - it will be so easy to know who we are because we have understood what we wanted.

I said at the beginning of this writing that mainstream

agriculture is fostering cultural suicide. It is a circle. The identity crisis (not knowing who we are), confuses what we want. One aspect of our mixed cultural history has been a vital foundation for our sensibilities and sense of place (in time). That aspect is our agrarian heritage - those centuries of ancestral agriculture deeply entrenched in human labor with sometimes accomplishment, loss, satisfaction and regret, but always a sense of home. Our new industrialized agriculture denies that heritage - that foundation - as it reduces and ridicules humanity in agriculture. So as we lose our heritage we stand the chance of collectively losing hope for our culture (if a culture can lose hope).

Human-scale, responsible, permanent agriculture should not be preserved just because we need or want to keep our heritage alive. It must be preserved for the sake of our cultural vitality. It must be not only preserved but fostered because within it lies the only hope for feeding the world, maintaining a healthy planet, and giving cultures a "reason" to continue. It is a circle.

As the small farm is replaced with agribusiness the transition bleeds away the substance of culture and purpose slips away. We can kill agribusiness but not by fighting it. That would be an exercise in shadow boxing because we'd be fighting ourselves.

Our offense is the key to our survival. But it is a mistake to fight in offense. Our offense must be construction. We must build what we want and ignore industrial agriculture for the misplaced misbegot cultural mistake it is. Also, however, part of our offense should be the preparation of defense. We must be intelligent in our efforts to defend what we build.

Go build a community of small farms and a radiant culture.

Chapter Eleven

Belonging to a Dream

There are many readers of *Small Farmer's Journal* who do not live with cows or ducks or rows of vegetables or fruit orchards or chickens or work horses. They live with the "someday" dream of those things. I feel a strong empathy with these dreamers. I know, first-hand, the pervasive wonder that comes with their imaginings. A wonder that includes "what if's," "if only's," and "will it ever's." A wonder that elates and depresses, but most of all persists. The dream, out of necessity, leads to practical questioning. And at that point the strength and character of the dream becomes threatened. It doesn't need to be.

I am thirty six years old. I grew up in suburbia and in the city. When I was five years old, riding the pinto rocking horse my father made, I dreamt of cows and calves and bulls and open spaces. I don't remember the source of the images. But I remember the cowboy daydreams. I remember that when I was ten, I used to walk to an older couple's home just to gaze across

Originally appeared in Small Farmer's Journal in the Fall of 1983

the yard fence. They had no lawn, just rows of vegetables and fruits and little fruit trees garnished below by colorful patterns of flowers. It seemed no square inch was wanting. Every piece of that large yard was part of some regular careful "doing." There were little paths with boards to walk on and there were funny tall handmade sprinklers and ornate, but rough, bean trellises punctuating a world of ordered foliage. And the wiry old couple seemed so happy - always so happy. I was a moody kid and at that time in my life sitting on the fire hydrant and looking into that oasis garden yard was a wonderful escape. I did not come to understand why until recently. I've never talked about it till just now. It was one of those private experiences that seem either embarrassing or insignificant. In truth there are answers to difficult questions well hidden in the dreams which draw us in and nourish us sometimes without apparent meaning.

Fifteen years ago I read a prose line in a book, by the French writer Albert Camus, which had stayed with me even though its meaning I had never grasped. He said <u>Love is that which in whose presence our hearts first open</u>. My heart first opened when I was astride my pinto rocking horse and adrift in my daydream. My heart first opened in the presence of that long ago yard garden. Yes, and since then my heart has "first opened" too many times to count. Love is that rocking horse and that yard garden. I have come to know what Camus meant. But I did not get there unassisted.

Almost two years ago my heart opened in the presence of a young woman. From the first she was able to see the things that mattered to me. Some of those things I had pushed away. She showed them to me again and in new ways. She told me that to deny my heart was a terrible death and she helped me to feel that. So I am getting reacquainted with my "heart." I am getting to know myself. I am looking forward and back and it's helping to make today richer.

In looking back I am seeing for the first time what it was that buried my dreams. I am writing this because I feel there may be value in hearing this personal history. I would like to think my words might help you hold a dream a little longer or perhaps return to a dream you have forsaken.

As a teenager, those experiences of the heart led me to know that the world I wanted to live in was a little farm. In no time at all, I found that I had best keep quiet about my dream. People my age and older, let me know that my goal was ridiculous, impractical and silly. I had to shoot for a career, I was told. Well, I succumbed at least physically, if not spiritually, and went through the hoops to end up an art instructor. But my dream wouldn't quit me. It engulfed me. In any spare moment I did things to feed the dream. I was always drawing little sketches of barns, sheds and farm houses. I drew plot maps of farm field plans with everything figured in, including where the rhubarb went. I read books, asked careful dumb questions and waited. The waiting was eased by vegetable gardening and chickens. I was attending college by day, bartending for a living by night and growing a garden and a dream all during my twenty first year.

When I moved to Oregon to do graduate work, it was not

because I had selected the best school, but because I had chosen
an area in which to build my dream. I rented a 25 acre farm
while working towards a Masters degree. I started in with
Nubian goats, Rhode Island Red chickens, ducks and ambitious
gardening. I worked at all sorts of jobs, but most often they were
in commercial agriculture. When I found time, I attended school.
My dream devoured me. It was no longer possible to hide my
ambitions. Anyone who came in contact with me knew that my
goal was to have a farm of my own. And people had lots to say
about that. There was some encouragement, but for the most
part there was concerned advice and criticism. I got an endless
stream of admonishments telling me all the things I could not do.

The total environment in which my dream had to live
began at some point to affect the shape of my goals. Under the
often inappropriate heading of practicality came all sorts of good
sounding reasons why I should try to get into mainstream agriculture. I was saved from that pressure quite by accident. In order
to survive, I had to work and the only work I could stomach was
in agriculture. The only places that could afford to hire someone
at today's inflated wages were large commercial operations. Four
years of working in such places convinced me by hard experience that large-scale industrialized agriculture was a dead-end
street economically and spiritually.

All along I had accepted the common notion that the only
way I could get started farming at any scale was with great
quantities of money. My only source of income was wages. At
some point I discovered that there was another way to buy a

farm. That was to borrow, buy, create phony equities, borrow more and buy more. I took the bait, not particularly for the farming, but to set up *Small Farmer's Journal.* It worked, but terrible prices in dollars and personal misery were paid. And I continue to pay.

I came to realize my dream and it was better than I hoped. Then I lost the dream. The dream is still very much alive for me, perhaps more than ever, but I do not live it now. It's been a couple years since I've farmed. I live on a farm and have some horses, cows and sundry stock. But the horses aren't working. I've been driving 200 miles round trip to the office to keep the Journal going and to raise the money to rebuild the dream. Still caught in the old trap. But all along, I've known better and been unwilling to see the truth.

My happiest moments, if I can be honest with myself, were on that first rented 25 acre farm. I checked and it's still there, rented by some people who took over after me. That 25 acre farm with a beautiful red brick house, barn, chicken house and cherry orchard rented for $125 per month. And the neighboring fields were all available for rent. In fact, most of it could have been purchased reasonably if a tenant could but prove worth by the care put into the place. But my suspicions, my ambitions, my greed, the distortions of my dream blinded me to the opportunity. I moved on and I'm still moving.

But there is a difference now. I can admit these things and I've rediscovered the simple purity of my dream. I know that it is the work, and the immediate world it creates and sustains, that satisfies me. Working horses I have raised in ground I care for, to raise crops that keep a cycle of growing, in an intertwined little environment with poultry and garden and livestock and everything else that fits - that is my place - my dream.

Always, when someone discovers you are a farmer, they ask "how big is your place?" Instead of the hollow answer - 150 acres - I want to answer with my dream in hand and say: "The land I work with extends so far as I can properly care for and love. It is as big and as small as I am. And this place does not belong to me, I belong to it."

Yes, all the questions about being able to pay the bills and providing adequately for the homestead are important and must be kept well in hand. But be careful with what you let the bills become. And look hard at all the things that are necessary and unnecessary in making a household adequate.

And that small farm you dream about can never be bought. It must be built to be real for you. To build does not necessarily mean constructing house, barn and fences. Although that may be part of it. It means working with and for. Think about renting, leasing or buying a little place with some of the basic ingredients and look to see that there is room to grow in concentric circles. An old horsefarmer friend owns five acres with a house and sheds. He rents and sharecrops 40 to 60 additional acres and is happy as a clam. (So are his landlords because he cares about the land.)

Having your dream come true does not mean possessing the dream. And your dream can come true, but only if you never lose sight of it. Keep it alive and protect its shape. You already know that having the dream within sustains you. I want to tell you that living the dream is also sustaining and more so. I want to tell you that the dream is good and it's worthwhile. Stay with it. It is possible.

You love that dream, it's opened your heart. It is up to you to bring the dream to life. When you do that, don't forget to use your heart as well as your mind and hands.

Your "someday" dream is your strength. That strength multiplied by many thousands of other folks means a lot of good things to many more people. I'm glad you're out there.